California High:
Warbirds of the West Coast

Michael O'Leary

Windrow & Greene

© 1992 Windrow & Greene Ltd.

Published in Great Britain by
Windrow & Greene Ltd.
5 Gerrard Street
London W1V 7LJ

Published in the USA by
Specialty Press Publishers &
Wholesalers Inc.
P.O. Box 338
Stillwater, MN 55082
(612) 430-2210/800-888-9653

A CIP catalogue record for this book
is available from the British Library

ISBN 1-872004-37-7

(Front cover) Curtiss P-40K (USAAF s/n
42-9749) was discovered near Port Hayden,
Alaska, where it had crash-landed on the
tundra during an operational World War 2
mission. The airframe was recovered for
restoration during the I970s, finally ending
up with Judge William Clark as N293FR.
Seldom flown, the fighter was sold at the
1990 Museum of Flying Auction, setting a
new record price for a P-40. Here the
macabre nose art of 85th Fighter
Squadron, 80th Fighter Group over Burma,
1943, is displayed to advantage by the late
Fred Sebby.

(Title page) One of the top three air racers
in terms of highest achieved speed is Tiger
Destefani's Mustang N71FT *Strega*.
Powered by a highly-modified Dwight
Thorn Merlin, the aircraft has set several
speed records at Reno, and constantly
duels with Lyle Shelton's Bearcat for first
place.

John Putnam maintains an impeccable formation in Wiley Sanders' beautiful *Georgia Mae*, one of two Mustangs owned by Sanders which divide their time between Bakersfield, California, and Troy, Alabama. NX10607 formerly saw service with the Royal Canadian Air Force as RCAF 9277 (USAAF s/n 44-74466) before being sold surplus in the early 1960s. In 1967 the fighter was purchased by John M. Sliker and used as a fairly stock Reno air racer. After Sliker's untimely death in a Grumman Bearcat accident the aircraft was purchased by Sanders. NX10607 has clipped wings and horizontal tail surfaces for air racing and, equipped with a full-race Zeuschel Merlin, is extremely competitive despite having none of the major modifications fitted to the top few racing Mustangs. Several years ago NX10607 was nearly written off in a landing accident at Reno, but Sanders had the damaged fighter lovingly rebuilt, including complete reskinning with new aluminium which has been polished until it almost glows.

inal victory in World War 2 left the United States with a huge aerial armada that was, for the most part, unwanted. Giant storage yards were set up across the country and thousands of wartime aircraft were parked in neat rows awaiting disposal. To the interested observer many of these aircraft, decorated with fabulous markings and nose art, characterised the American spirit during combat. To the uninterested, the machines were merely war surplus metal waiting to be turned into items that would benefit the taxpayer.

Several of these vast storage yards were in California; and, during 1946, if you happened to be driving down Euclid Avenue in Chino past the farms and dairies, you would pass Ontario Army Air Field and, perhaps, notice the hundreds of aircraft parked in orderly rows, their national insignia daubed out with drab paint, the weeds growing up past their wheel rims. Ontario AAF, site of today's Chino Airport, was stuffed with a variety of aircraft – everything from war-weary Grumman J2F Duck biplanes to brand-new North American B-25 Mitchells flown directly from the Inglewood factory to the field. Very few of these aircraft would survive the next two years.

So what do you do with an air force that no one wants? With storage yards full of unwanted aircraft, the Reconstruction Finance Corporation began to dispose of the machines. Many trainers were sold to private pilots who wanted a cheap aircraft, but the bombers and fighters mainly went to the scrapper. (Any viewer of the excellent film 'The Best Years of Our Lives' will certainly remember an unemployed Dana Andrews walking among rows of abandoned Flying Fortresses, wearing his old A-2 flying jacket and remembering his days of combat over Europe; the scene was filmed at Chino.) Although most of the bombers and fighters met their end in smelters, some did survive, either to be sold to foreign air forces, held in storage for the government, or sold to individuals who wanted a unique method of transportation.

With the unexpected outbreak of the Korean War the American government found itself desperately in need of combat aircraft, and many veteran prop-driven warplanes were pulled from storage and recon-ditioned for military service. The California Air National Guard had been equipped with Mustangs, and many of these were also sent to Korea to aid the beleaguered Allied forces. The odd situation developed where American civilians were operating a few privately owned Mustangs in California while their countrymen in uniform were getting seriously shot at in the very same aircraft in the skies over Korea. The Mustangs were now mainly employed for ground attack, and performed this mission admirably, even though the coolant system for the V-12 Rolls-Royce/Packard Merlin was extremely vulnerable to enemy ground fire. The sudden appearance of the Chinese MiG-15 *Fagot* jet made life even more difficult for pilots operating the World War 2 piston-engined fighter.

Back in California, the Grand Central Air Terminal in Glendale was responsible for bringing many Mustangs back into combat shape for Korea, as well as preparing surplus P-51s for Third World air forces – particularly in Latin America. Grand Central Air Terminal had been the main southern California airport during the 1930s and was known as 'the airport of the stars' because of its close proximity to Hollywood. Until it was replaced by Los Angeles International, Glendale was the airport of choice for Los Angeles; but during the 1950s its ramp was lined with hundreds of surplus fighters and bombers being reworked for new combat missions.

After Korea, the California ANG kept their Mustangs until they received newer equipment in the form of North American F-86 Sabres and the Mustangs were shuffled off to disposal. The last Mustang surplus sales were held in California during the late 1950s, and the aircraft were quickly snapped up by civilians – some to be flown again, some to be sold down south and some to be cannibalised for parts. Aircraft prices at the McClellan AFB auction ranged from $1,500 to $2,500. With the military equipment ripped out and the old radios and fuel tank behind the pilot removed to make room for a second seat, the surplus Mustang represented the ultimate aerial 'hotrod'.

California was also home during this period to

Over a lake in the Sierra Nevada foothills, two Mustangs practice some formation flying. The lead aircraft is N1451D owned by Mike Clark, while the wingman is the late Rob Satterfield flying N7722C (USAAF s/n 44-73420). The photo illustrates the difference between a polished metal P-51 (background) and a bare aluminium Mustang. Currently, most owners favour having their warbirds painted silver since the paint helps protect the aluminium skin, but nothing beats the look of a warbird that has had the aluminium buffed to a deep natural glow.

several eccentric collectors who purchased large numbers of surplus aircraft and hoarded them. One was a unique character named Bob Bean, who gathered a large collection of Corsairs and Lightnings in California and Arizona. Fortunately the aircraft were not scrapped, and as the values for such machines picked up in the 1970s Bean began to sell off items from his fleet to restorers.

During the 1960s it was not uncommon to see a variety of former fighters tied down on the ramp at many California airports – everything from P-51s painted in a variety of garish civil colour schemes, to the sole surviving North American P-64 rotting on the edge of a small field near Riverside. Little was done to maintain them, and if something broke the aircraft were often just parked until the owners figured out what to do.

Although prices did not really increase all that much during the 1960s, a new group of buyers appeared. California during the 1950s had been the 'hotrod' and drag-racing capital of the world, and sometimes the drivers and builders of these cars would gravitate to the surplus fighters, attracted by their promise of speed and power. Dave Zeuschel and Frank Sanders, well known in the drag-racing field, learned to fly and subsequently bought Mustangs, which led to purchases

of similar types of aircraft by other former 'hotrodders'.

By the early 1970s the ex-military machines had become known as 'warbirds', and workshops had opened at various airports around the state specialising in the rebuilding of these aircraft and their engines. Airshows, such as the annual 'Gathering of the Warbirds' at Madera, became showcases for the restored aircraft, and public interest began to increase dramatically – as did the prices asked.

From that point on there was really no looking back, and ex-military fighters of all types poured into the state for restoration and flying. During the 1980s it was not uncommon to see almost 30 Mustangs at some airshows, along with a smattering of the rarer fighters like Thunderbolts and Lightnings. The restoration even of vintage jets became popular, and with the importation of ex-Chinese MiG-15s the situation really got interesting.

Warbird fighters always draw a crowd wherever they go, and the aircraft have become a revered part of aviation's heritage. The acquisition and restoration of these fine fighters continues unabated, and there is little doubt that the 'sky sharks' will be operating for many years over California.

Californian airshows attract warbirds from across the nation, allowing pilots to get together and enjoy their machines. This particularly nice example of North America's most famous fighter is registered NL51KD and is owned and flown by Dean Cutishall. Originally built as USAAF s/n 44-73436, the aircraft was sold to the RCAF where it flew as 9300. The fighter went through several owners and registrations (N6313T, N51TK) before being acquired by Dean, who regularly attends airshows and has raced at Reno in NL51KD. Here it shows off the attractive livery of the 8th Army Air Force's 375th FS, 36lst FG.

(left) Wiley Sanders' second Mustang is camouflaged N51WB which is of historic significance since it was the first Mustang built in Australia (designated CA-locally) under license by the Commonwealth Aircraft Corporation. Serial 68-1 was delivered to the Royal Australian Air Force in June 1945 and was struck off charge in 1953. Along with A68-30, -35, -72 and -87, the Mustang was then staked out on the desert floor near the Woomera Rocket Range as part of the British atomic bomb test programme. Placed half a mile from

the blast tower, the aircraft were subjected to two detonations, the effects being carefully monitored by military personnel. Fortunately the aircraft emerged with only minor damage and, after repairs, was flown out of the test area in October 1967. The P-51 was damaged during transit to America in 1969, and its condition did not improve as it passed through several owners, before being completely restored by Don and Bill Whittington during the early 1980s. This aircraft also features clipped wing tips for air racing.

The grand master of Mustang flying is Bob Hoover, whose career has encompassed flying Spitfires with the USAAF (he was shot down and captured) and an illustrious period of test flying with North American. However, N51RH is arguably the most famous of all Mustangs, and Bob's yearly start of the Reno unlimited air races was only recently eliminated because the stock Mustang was too slow as a pace aircraft for the new generation of racers! As with many of the surviving Mustangs, USAAF s/n 44-74739 went on to serve with the RCAF (as 9297), and other civilian identities have included N8 672E and N151Q.

(Above) Certainly one of the most attractive Mustangs currently flying is Peter McManus' N51PT which is finished in the personal markings of John C. Meyer, one of the higher-ranking 8th Air Force Mustang aces. NL51PT was built as P-51D-20-NT USAAF s/n 44-72145 and has carried the previous civil identities N311G and N6169C. McManus performs a superb aerobatic act that really shows off the Mustang's potential.

(Left) During the 1981 Reno air races, CASA HA-1112 Buchon (the Spanish-built variant of the Messerschmitt Bf 109, fitted with a Rolls-Royce Merlin) N109DW was damaged following a ground loop on landing – the 109's characteristic vice. The owner donated it to The Air Museum, who placed the fighter in storage until a decision was taken to make it airworthy during 1989. A considerable amount of repair work was needed; and while this was being undertaken the staff decided to modify the aircraft's nose contours to make them more closely resemble the lines of the original Daimler-Benz variant (the Merlin cowling is, admittedly, unattractive). A new lower cowling and spinner were built up from fibreglass and aluminium to create a more nearly authentic appearance. Repainted in a spurious *Luftwaffe* scheme and re-registered NX700E, the HA-1112 made its first flight during May of that year; here it is being flown by John Maloney.

Dago Red, NX541OV, flown by Bruce Lockwood, shows how highly a Mustang can be modified for air racing. Clipped wings, cut down canopy and smaller radiator scoop are just some of the mods that made *Dago* one of the faster racers around the Reno pylons during the mid-1980s. Newer aircraft have now bumped *Dago* out of the top six speedsters, and further modification work will be required to make it a front runner again. In the background is Bob Pond's N151BP in the 375th Fighter Squadron's blue finish.

(Above) Pioneer Aero has remanufactured three Mustangs for British collector Doug Arnold and the first of these, registered NL314BG and also finished in one of John C.Meyer's schemes, is seen during its first flight in July 1988 with Elmer Ward at the controls. The aircraft was subsequently ferried to Britain by Mike Wright.

(Below) Southern California, as well as being the base for many warbirds, is also home to Hollywood, so it is not unnatural for the two forces to combine when aircraft are needed for films or TV. During filming of the pot-boiler 'Hindenburg' some German aircraft were needed for set dressing, so The Air Museum's ultra-rare (but non-flyable) Messerschmitt Bf 109G-10/U-4 *werke nummer* 611943 was trucked to Camarillo Airport, which was standing in for a German field. A flyable four-seater

Bf 108 was also utilised, and the Bf 109 was painted in a wholly anachronistic Russian Front JG54 'Green Hearts' scheme to match the already-painted Bf 108. The Air Museum's Gustav, tested post-war by the United States and given the evaluation number T2-122, was one of the aircraft rescued from imminent scrapping by Ed Maloney in the 1950s. There has been some discussion at the museum about making this aircraft flyable.

ove) One of two dozen ex-Iraqi Air
ce Hawker Sea Fury FB.10s imported to
United States during 1979 by David
ichet and Ed Jurist, this aircraft
gistered N4OSF at that time) was sold
a restoration package to Guido Zuccoli
ng with several other Furies) and
pped to Australia by sea. Later sold to
Allen, the Fury was restored to flying
dition as VH-HFA with the apt name
gnificent Obsession. In 1988 the Fury

headed back to the States and arrived at
Chino, where it was licensed by Pacific
Fighters as NX57JB for new owner John
McGuire. John Muzsala is seen here over
Chino during a test flight in the newly
assembled aircraft.

(Below) A flyable German aircraft was
needed for a brief sequence in
'Hindenburg' when George C.Scott's
character supposedly returned to Germany
from service with the Condor Legion in
Spain. The producers rented this HA-112
registered N109ME (Spanish serial C4K-
31), and Gerald Martin flew it from Texas to
California. It had already been painted in
Condor Legion Bf 109E markings, so the
film had a curious mix of 109 schemes.

One of two Sea Furies in the Frank Sanders stable is N924G, a T.Mk.20S that was operated by *Deutscher Luftfahrt-Beratungsdienst* as D-CAMI after being surplused from Royal Navy service (VX300). Sold in Britain during 1974, the aircraft became G-BCKG with Warbirds of Great Britain, but was soon sold to John Stokes in Texas as N62147. The aircraft was obtained by Frank Sanders during 1978 and completely reworked to his specifications. Few who saw Frank's airshow display in this machine, with the wingtip Smokewinders belching out a white trail, will ever forget the performance. In this photograph the Sea Fury is being flown by Dennis Sanders.

As mentioned in the text, California became home to several aircraft collections owned by rather colourful individuals. This May 1968 photo shows just a portion of the Bob Bean collection at Blythe, California. The three P-38s and one Corsair seen here were all eventually disposed of by Bean. The extremely rare F-5G photo-recon Lightning went to the Pima Air Museum in Arizona before transferring in the late 1980s to the *Musée de l'Air* in Paris, where it was destroyed in a disastrous hangar fire in 1990. The other two Lightnings (ex-Honduran Air Force) and the Corsair were sold to collectors and made airworthy.

(Above) Bill Rheinschild owns Mustang N35FF named – appropriately for the world of unlimited air racing – *Risky Business*. The aircraft features clipped wings and a number of other racing modifications, including a very hot Merlin, and has been quite successful at Reno. In the photo 'Race 45' is being flown by Matt Jackson.

(Left and below) Rick Brickert poses the attractive 'Flying Tiger' Curtiss TP-40N Warhawk acquired in 1980 by Bob Pond's Planes of Fame East in Spring Park, Minnesota. During the winter Pond bases some of the aircraft at Chino and Palm Springs, where they are regularly flown. USAAF s/n 44-7084 had been displayed for many years in the USAF Museum before owner Chuck Doyle took the aircraft back and registered it as N999CD. This aircraft was a P-40N that had been converted by the factory into a dual-control TP-40N training variant. Doyle dispensed with the TP's awkward double canopy arrangement and installed a unit that looked more like an original N-model rear canopy; the two seats were, however, retained.

(Right) One of the most flown of all the Curtiss fighters is the P-40M Kittyhawk N1232N that was owned, until 1992, by Tiger Destefani. Originally built as USAAF s/n 43-5795, the aircraft went on to fly with the RCAF as 845. In 1947 it was one of a large group of P-40s sold surplus in the States and was registered N1232N. Over the years the aircraft led a rough life, even being employed as a cloud seeder before being abandoned in Sonoma, California, during the early 1960s. In 1964 N1232N was purchased by Harrah's Auto Collection in Reno and put on static display. When the museum got rid of its few aircraft during the early 1980s the fighter was purchased by Tiger Destefani in Bakersfield, California, and completely restored. Fred Sebby is seen practising some close formation work. This aircraft, like most P-40s, has been converted to house an extra seat behind the pilot.

Razorbacks at their best. Ray Stutsman's Curtiss-built Republic P-47G Thunderbolt, painted in the markings of Captain W.C.Beckman's *Little Demon*, is kept company by Pete Regina's North American P-51B Mustang finished as Don Gentile's *Shangri La*. Stutsman found the Thunderbolt in a Los Angeles junk yard and completely restored the fighter at his shop in Indiana. Stutsman campaigned N47DG for several seasons before selling the 'Jug' to the Lone Star Flight Museum in Galveston, Texas, where the fighter is kept airworthy. Regina's B-model Mustang, N51PR, was rebuilt from the ground up, using an original B wing as the basis of the restoration. This aircraft is now owned by Joe Kasparoff and finished in a shocking overall red paint scheme.

During the 1950s and 1960s the value of warbird-type aircraft was limited, and it was not unusual to see them in disrepair or simply abandoned. Bell P-63A Kingcobra NX90805 landed at Van Nuys in the late 1940s, straight from a surplus sale, and never flew again; its condition went downhill rapidly, and it was vandalised a number of times. During the 1970s the Kingcobra was pushed into a hangar where some minor restoration work was undertaken, and it has remained there ever since, the owner refusing to part with the rare mid-engine Bell fighter.

The late Dave Zeuschel is seen flight testing P-47D N47DE, USAAF s/n 45-49025, near Chino prior to its ferrying to Britain by Mike Wright. It is one of six P-47Ds recovered by Vintage Aircraft International and shipped in 1969 to Texas, where their arrival was a great boost to the growing warbird movement. While in Britain the aircraft was rarely flown, and it was acquired by Stephen Grey in 1985 as G-BLZW. It was not long before the fighter was sold to Bob Pond and rebuilt by Steve Hinton at Chino as N47RP. The huge non-standard underwing drop tanks aided the thirsty R-2800 on its transatlantic crossing.

(Below) The operation of any high-performance equipment brings with it a certain amount of danger. The Air Museum's P-47G, N3395G, suffered an engine failure while performing at the 1971 NAS Point Mugu airshow. The pilot managed to get the stricken fighter down in a farmer's field, but extensive damage resulted. Happily, the Thunderbolt would fly again.

2: Basically Bombers

One of the most significant warbird restorations of the early 1990s is David Tallichet's magnificent Martin B-26 Marauder. This was no short-term project, since Tallichet (and crew) recovered three Marauders that had been forced down in Canada on a 1942 ferry flight way back in the early 1970s. Although all suffered

some form of damage they were in remarkable condition, having been left basically intact with all the original military equipment still inside; the cold climate had kept the B-26s in good shape and corrosion was minimal. A first flight was scheduled on 5 February 1992 from the aircraft's home base at Chino, and a large

contingent of Marauder veterans were on hand – including some who were aboard the three aircraft owned by Tallichet – but an engine malfunction while running up the Pratt & Whitney R-2800s aborted the planned event. The three aircraft recovered were 40-1451, 40-1459, and 40-1464; it was the last that was chosen for restoration.

Seen during March 1977 at Chino, the remarkable condition of the Marauder is evident – its paintwork only started to deteriorate after arriving in smoggy Chino. The B-26s obtained by Tallichet were from the first batch of Marauders built for the military, and have the infamous 'short' wing that led to the aircraft possessing an interesting engine-out performance.

Why would anybody in their right mind try to keep a veteran bomber in flying shape? The proposition would, from the start, seem to be a contest with no winner. The fact that California is home to nearly 50 such aircraft means that there are a lot of folks in the state who *want* to keep the things in the air, and damn the cost.

After the war there were a number of specialised civilian uses for ex-military bombing aircraft. A few Boeing B-17 Flying Fortresses went to companies which needed a high-altitude mapping/photographic platform; the Fort was ideal for such work, with its turbocharged Wright engines and large load-carrying ability. Several B-17s were converted into business aircraft; a particularly nice example was operated for several years by the Chicago *Tribune* newspaper, while another was completely rebuilt by Boeing for TWA and was later passed on to the Shah of Iran as a sort of 'Air Force One', courtesy of the Central Intelligence Agency. Boeing, eager to gain new contracts in the cost-conscious years following the war, even went as far as drawing up plans for a 'boardroom bomber' – an executive conversion that would offer a number of upgrades and a comfortable cabin for long-distance flights. However, interest was apparently minimal and Boeing quietly dropped the project.

New jet-age bombers were rapidly replacing the war veterans still in service, but the military continued to find gainful employment for some of the remaining aircraft. The tough Douglas A/B-26 Invader saw much use during the Korean War, where its load-carrying abilities and heavy armament were much appreciated. B-17s found new missions as VIP transports and as target drones and directors. B-25 Mitchells became sturdy crew and radar trainers. Other aircraft such as the Martin B-26 Marauder and Consolidated B-24 Liberator were simply deemed to be of no post-war use, and were scrapped in large numbers.

Civilians visiting the big storage yards picked up low time surplus Mitchells and some of the few Invaders offered after the war. One or two Invaders were set up as racing aircraft, and a few were converted as high speed executive transports, a sign of what was to come in the late 1950s. Mitchells were snapped up for a number of general duties including cargo hauling while Navy bombers like the Avenger and Harpoon found ready buyers for a variety of workaday tasks.

Fire-bombers

Massive home building had taken place in California after World War 2 and many of the once barren hillsides now sprouted huge housing tracts. This posed a problem, since southern California has a long fire season in which tinder-dry hillside (and houses) can quickly become raging infernos. With the rapid increase in population, it became apparent that

▶

After the Marauders had been disassembled in the Canadian wilderness they were helo-lifted to Fort Smith and then sent by train to southern California. Tallichet had this aircraft finished in the same Olive Drab and Neutral Grey colour scheme in which it was found. The quality of workmanship, and the power turret, are clearly seen in this view.

something had to be done to combat this threat to property and lives. The answer lay within the surplus fleet of wartime bombers. By rigging tanks in former bomb bays the aircraft could be converted into a fleet of aerial tankers that could quickly respond to any fire threat and bomb the area with fire-retardant borate. Several Avengers, Mitchells and Forts were so converted and pressed into service, becoming an almost immediate success.

The effectiveness of the fire-bombers led to the creation of a number of companies within the state specialising in this mission, and surplus Avengers, Mitchells, Forts and Privateers were snapped up for conversion. Most of these businesses were real cowboy outfits, the owners and pilots alike being mavericks who enjoyed flying ex-military equipment in an almost war-like environment. Needless to say, as in combat, there were casualties: during the early 1960s several B-25s suffered massive in-flight structural failures while fighting fires, and the type was eventually banned from such operations. Similarly, crashes of Grumman Avengers while fire-bombing led to a ban on single-engine types during the early 1970s.

The big aircraft like the Forts and Privateers soldiered on against man's oldest enemy. By this time aircraft were beginning to show their age, and most Fortresses were withdrawn by the mid-1980s; but the five remaining Privateers still venture forth to do battle with the flames from their base with Hawkins & Powers at Greybull, Wyoming, and show little sign of retiring. The former bombers have been replaced by newer aircraft like the Tracker, Neptune, Hercules, and a variety of Douglas transports from the DC-4 to the DC-7.

'Catch 22'

The grounding or disposal of the World War 2 veterans meant that many new aircraft were on the market, but when these machines were disposed of in the late 1960s and early 1970s few went to collectors. Most of the Avengers went to Canada to be converted into sprayers to fight the Spruce Bud Worm. The B-25s simply sat at their former operational fields as weeds sprouted around landing gear legs; but the late-1960s film 'Catch 22' saved many of them from a date with the scrapper. The film company contracted with Frank Tallman at Orange County Airport (before it received the rather awful name of John Wayne Airport) to create a squadron of combat-ready Mitchells for the film.

Veteran aviator Tallman scoured the fire-bomber bases and pulled together a diverse collection of marginally airworthy Mitchells, which were ferried to Orange County for urgently needed maintenance and the addition of military gear such as turrets and paint schemes. The film was (deservedly, in the author's opinion) a flop; but 18 Mitchells had been saved, and

Tom Reilly's 'Bombertown' in Kissimmie, Florida, has completed many magnificent restorations; but the company's crowning glory has to be Consolidated B-24J Liberator (USAAF s/n 44-44052) N224J, nicknamed *All American*, which made its first flight during September 1980 after an exhaustive rebuild for the Collings Foundation. The aircraft is seen over Palm Springs during September 1989 as it headed for San Diego to celebrate the 50th anniversary of the first flight of the type. The names of the numerous donors and sponsors who have contributed to this massive undertaking cover the fuselage.

these were put up for sale by Paramount after the movie's completion and sold for between $4,000 and $6,000. Even at those prices it took a long time to dispose of the aircraft; but these Mitchells would go on to form the basis of today's finely restored B-25 fleet.

Douglas Invaders were desirable aircraft since they could perform a variety of missions extremely well. After the Korean War more Invaders became available for the surplus market, and most were snapped up by companies for conversion into high-speed executive transports. The company that finally came out on top was On-Mark, based at Van Nuys, California. During the late 1950s and early 1960s the On-Mark hangar was

busy churning out customised Invaders in a variety of variants (including a fully pressurised example) for the business market. As the Vietnam War escalated, On-Mark built the Counter-Invader for the USAF. This highly modified Invader featured upgraded engines, avionics, a strengthened wing and a greatly increased combat capability. The aircraft was heavily employed in South-East Asia, but after the war the surviving examples were flown to Davis-Monthan AFB for scrapping, with specific orders that they not be released to civilians.

Perhaps the most classic of all wartime bombers is the B-17 Flying Fortress. Vintage aircraft collectors

David Tallichet picked up his G-model Fort after it had been retired from fire-bombing duties. Tallichet had served as a co-pilot with the 'Bloody Hundredth' (100th Bombardment Group, 8th Army Air Force) based at Thorpes Abott during World War 2, and has always maintained a soft spot for the four-engined warrior. Based at Chino, Tallichet regularly flies his aircraft to airshows, and the B-17 became the lead Fort in the 1990 film 'Memphis Belle'. Currently, this is the only flying B-17 based in California.

Tallichet also owns an extremely rare Consolidated B-24J Liberator, obtained from the Indian Air Force when that service phased the type out during the 1970s.

This splendid machine is one of only two B-24s still flying, and is an active participant in airshows nationwide.

Many B-25s, A-26s and TBMs still fly in California, and are prized by collectors who go to great lengths to restore the aircraft to original World War 2 specs. This involves trying to find items like turrets and interior equipment to bring their machines back to stock condition – no easy task. California is the 'bomber capital' of America, with a large percentage of the today's flying veterans based within the state, and several other airframes undergoing restoration by companies specialising in the task.

N224J was operated as Liberator B.VII KH191 by the Royal Air Force during the late war years. When hostilities ceased the airframe was simply left, along with hundreds of others, in India. In order to make the aircraft non-airworthy some damage was usually done to each airframe; but the inventive Indians had lots of time on their hands, and hundreds of aircraft from which to draw, and eventually they succeeded in forming two squadrons. The aircraft enjoyed a busy second career on coastal patrol, and the last example was not retired until the late 1960s. Note here some of the detail that went into the intensive Reilly rebuild, the bomber being virtually reskinned. The interior of N224J was completely refinished with original military equipment including power turrets and replica .50 calibre Brownings hooked to loaded ammunition belts. N224J had been relegated to the role of training airframe, serial T-18, at the Indian Air Force Technical College in Jalahalli after two decades of frontline service.

Banking steeply over the California desert, N224J shows off its chunky lines, and the markings of the 465th BG, l5th AAF. The aircraft was in really terrible condition when acquired by British collector Doug Arnold, with heavy corrosion and much of its interior gutted. Arnold had the aircraft dismantled and flown to Britain inside a four-engined Short Belfast owned by Heavylift, arriving on 6 May 1982. While in Arnold's ownership little was done to the bomber, which was stored at Blackbushe, except to have the fuselage polished. This view shows the aircraft's high-aspect ratio Davis wing to advantage. Currently, Bob Collings keeps the B-24J on an active airshow schedule, and the aircraft brings in a substantial amount of money which contributes to its yearly upkeep.

(Right) The only other stock Consolidated B-24J Liberator still flying is the example owned by David Tallichet. This B-24J (USAAF s/n 44-44272) also served with the RAF during World War 2 as KH401. As with *All American*, the Liberator was abandoned after the war in India, but was made airworthy and operated with serial HE771. The bomber was finally retired on 31 December 1968 and stored at Poona, where it was purchased by Tallichet in 1973. He registered it as N94459 and flew it from Poona to Duxford, England, arriving on 28 October 1973. N94459 spent a couple of years as a very welcome guest at historic Duxford before finally flying to the States in 1975. Since then the Liberator has been subjected to an intensive restoration and is now flown regularly on the airshow circuit as *Delectable Doris*, in the markings of the 566th BS, 389th BG, 8th AAF; it has made several joint appearances with *All American*.

During the 1950s and 1960s many stock Douglas A-26 Invader airframes were converted into high-speed executive transports, these modifications greatly altering the original configuration of the famous American bomber which fought in three major wars. When the warbird movement began in earnest in the early 1970s, few stock airframes were available for restoration to military status. One of the exceptions was A-26B USAAF s/n 41-39401, which had been stored for many years at Van Nuys. The aircraft had seen combat service in World War 2 and Korea and had been registered N3457G when surplused. Obtained by Challenge Publications in 1982 and registered N39401, the aircraft was brought back to flight status and a few missing military items were added. The resulting aircraft was a stock Invader, complete with operating turrets, which completed its first post-restoration flight on 18 August 1983. In 1987 *Whistler's Mother* was sold to the Weeks Air Museum in Florida.

One of the most impressive bomber restorations of the late 1980s was that carried out on Doug Lacey's magnificent Lockheed PV-2 Harpoon patrol bomber. The aircraft had been purchased as a tired and gutted fire ant sprayer, but Doug's dedication led to a very high quality

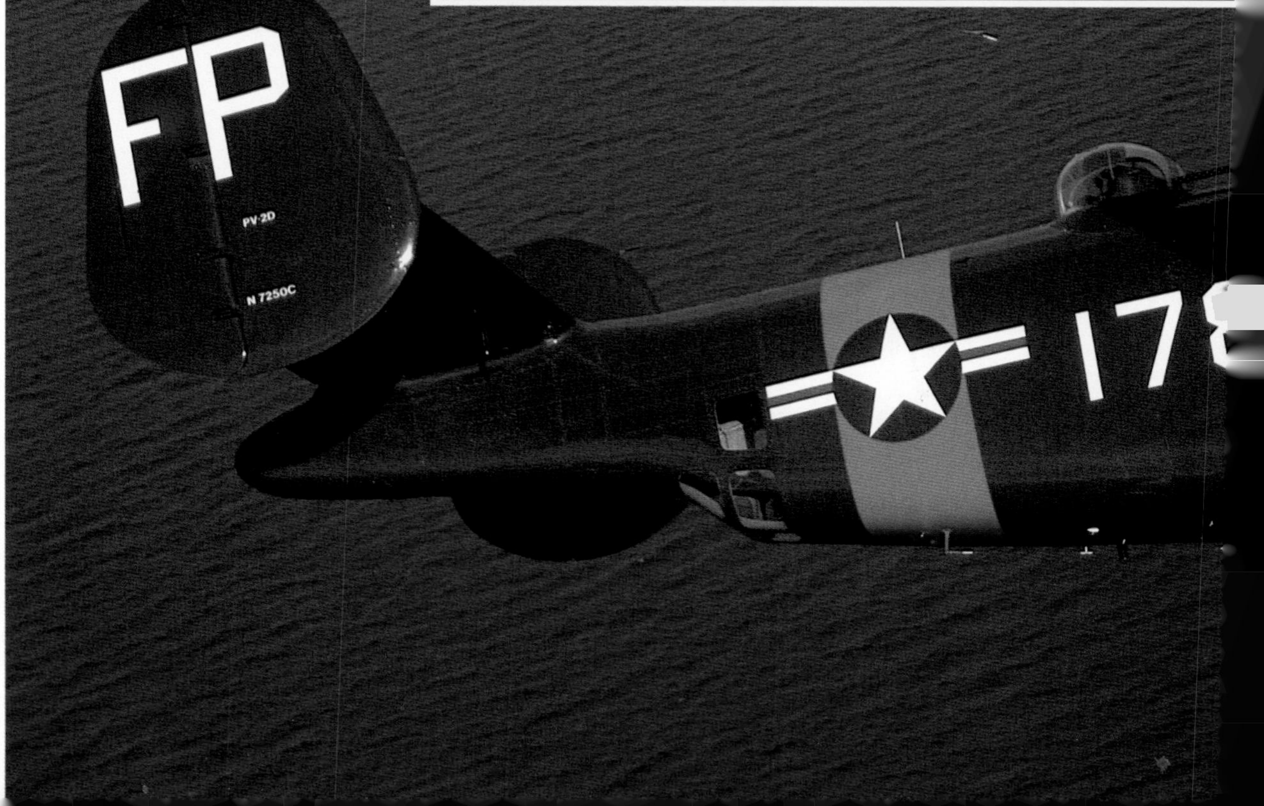

FP

PV-2D

N 7250C

…storation that saw all the original military …ear re-installed in the aircraft, most of …hich was in perfect working condition. …ne of the often overlooked combat …rcraft of World War 2, the Harpoon saw …eavy action in the Pacific. Lockheed built … total of 500; N7250C was a rare PV-2D which had an eight-gun nose fitted – Lacey restoring it in very accurate post-war US Naval Reserve markings. Banking steeply away from the camera ship, the PV-2D displays details such as the belly gun position, underwing rockets, and operable bomb bay fitted with 500 lb bombs.

Tragically, shortly after this photo-flight over Lake Tahoe in June 1990, while appearing in an airshow at Clearlake, California, the Harpoon stalled on the pull-up from one of a series of passes over the lake and plunged into the water, killing all eight on board.

TB-25N (USAAF s/n 44-30606) N201L in flight over Lake Tahoe. This particular machine had been surplused as N5249V in 1963 and was later one of the few Mitchells converted to executive configuration, with picture windows and comfortable interior. After being disposed of by its corporate owner the bomber's condition deteriorated; but in 1989 the TB-25 was purchased by Ted Melsheimer and, with the help of Mitchell specialists Aero Trader, was returned to fine flying condition – as can be seen in this June 1990 photograph.

(Below, left and right) *Heavenly Body* is the attractive Mitchell owned by Mike Pupich and based at Burbank. Featuring provocative Vargas-style artwork and the markings of the 42nd Bombardment Group as seen over Guadalcanal in 1943, the aircraft is TB-25N USAAF s/n 44-30748, registered N8195H. The Mitchell had been operated as a fire-bomber by Avery Aviation at Greybull, Wyoming, before being pushed off into the weeds – but then made airworthy once again for '*Catch 22*'. Pupich and his volunteer crew have lavished thousands of hours on the aircraft to make it airworthy once again.

One of the most successful of all attack aircraft is the hard-hitting Douglas Skyraider. Jay Cullum displays the rugged lines of his AD-4NA over Madera. Designed in 1944 as a contestant in the Navy's new single-seat BT category of attack aircraft, the Skyraider came about during an overnight brain-storming session as Ed Heinemann, Leo Devlin and Gene Root produced sketches in their hotel room. The all-night gathering proved its worth, as Douglas would go on to build 3,180 Skyraiders over a twelve-year period. Jay's aircraft, finished in the markings of VA-176 'Stingers', last saw active duty with the *Armée de l'Air*, and carries the civil registration N409Z.

(Right) Aero Trader co-owner Carl Scholl leads a pack of Mitchells in a formation flight at the annual Madera Gathering of Warbirds. The lead aircraft is TB-25N (USAAF s/n 43-28204) registered N9856C. Painted as a US Navy PBJ-1J, *Pacific Princess* is owned by Ted Itano and based at Chino.

(Below) The Beech AT-11 Kansan was a modification of the basic C-45 transport to turn the aircraft into a trainer for bombardiers, navigators and gunners; after the war surplused aircraft were used in the photo-mapping role. The AT-11 is currently a fairly rare machine. This very nice example, *Swamp Chicken*, registered C-GJCC and owned by Jerry Janes when photographed over Shafter, California, is being flown by Howard Pardue. The AT-11 could be fitted with a small power turret or a navigational blister.

(Right and below) Certainly the most magnificent of all four-engined bombers, about a dozen B-17s still fly. One of the nicest examples is the recently restored *Miss Museum of Flying*, which made its debut at the 1991 MOF Auction. The aircraft had been stored for about ten years before being obtained and rebuilt by Don Whittington in 1991. Registered N3509G, the B-17G (USAAF s/n 44-85778) had served for many years as a fire-bomber.

(Below right) Currently, one of the most popular airshow acts involves 'attacks' on Pearl Harbor by Japanese warplanes. The stars of the show are aircraft that were built for the 1968 epic '*Tora! Tora! Tora!* ', 20th Century Fox contracting to have a large number of T-6s and BT-13s converted to look-alike *Zeros*, *Vals*, and *Kates*. The results were quite realistic, and the aircraft were snapped up by private owners after the film. Two of the nicest are the *Zero* and *Val* owned by Gene Fisher, seen here over downtown Los Angeles during late 1991 with John Muszala flying the Val dive-bomber and Dennis Sanders in the *Zero*.

One of the most popular bombers on the airshow scene is the beautiful *Sentimental Journey*, owned and operated by the Arizona Wing of the Confederate Air Force. Registered N9323Z, the B-17G had spent many years with Aero Union in Chico, California, as a fire-bomber. Obtained by the CAF, the aircraft was in good flying shape but lacked any military equipment, and the Arizona Wing has spent years bringing the aircraft back to pristine condition. The bomber earns its keep by visiting airshows and is in California on a regular basis; the crew take sensible precautions against mishaps far from home.

3: Chino and The Air Museum

The Air Museum is a repository of many rare and historic aircraft, of which a good percentage are maintained in flyable condition. One of the rarest is the only fixed-wing Sikorsky still flying, the AT-12 Guardian. Originally constructed for the Swedish government as SEV-2PA-204A two-seat fighter bombers, 50 of these aircraft were ordered along with 100 EP-1 single-seat fighters similar to the P-35. However, the United States placed an embargo on military aircraft sales to Sweden, and aircraft not delivered were confiscated. Forty-eight SEV-2PA-240As became AT-12 advanced trainers and scattered to various training bases. After the war several AT-12s were sold surplus, and one even attempted to become a cross-country racer. N55539 spent some time in Latin America before it was rescued by Ed Maloney, and today remains a wonderful reminder of a time long past. Ross Diehl is seen flying the sole survivor near Chino.

Of all the airfields in California that are home to warbirds, the most famous is Chino Airport; in fact, if there was a Mecca for warbird pilots and enthusiasts then Chino would be the spot. Under its WW2 title of Ontario Army Air Field, Chino served as a primary training base – and even played a starring role in the Abbott and Costello potboiler 'Keep 'Em Flying'. Undoubtedly the best part of this rather predictable film was the impeccable stunt flying done by Paul Mantz in his rare Boeing 100.

For several years after the war Chino was the site of intensive scrapping of surplus warplanes, so it is ironic that the field today is one of the premier sites of warbird restoration. A good proportion of the airfield is used for agriculture; although a couple of crumbling bomb revetments still stand, the majority have been torn down over the past decade. As crops are harvested it is possible to walk over the fields and uncover jagged pieces of aluminium – all that is left of the vast aerial armada destroyed here in the late 1940s. (The author once unearthed a near-mint Air Transport Command enamel uniform badge in this way.) Although the future of warbirds is currently at Chino, the past is not very far away.

During the early 1950s a young man named Edward Maloney realised that the historic aircraft of World War 2 were rapidly disappearing, and that few people gave a damn. Although of limited resources, Maloney decided to do what he could to preserve our aviation heritage. Scouring scrapyards and technical schools, he was able to make a number of significant coups. On learning that the Navy was going to junk two Mitsubushi Zeros, Maloney rushed to the airfield and found that one aircraft had already been cut in two but that the other was still intact. In short order, a deal was made that saw the two fighters saved. When a technical school decided to get rid of its unwanted aircraft Maloney came away with a P-51A Mustang, a P-47G Thunderbolt, and an incredibly rare Messerschmitt Me 262 jet fighter.

Storing his rapidly growing collection became a major problem since the aircraft were easy targets for vandals. During this time Maloney did not really envision a flying museum, but he knew that unless he did what he could many historic aircraft would be lost. Collecting these aircraft did not always represent victories – Maloney actually lost more than he gained as thoughtless bureaucrats scrapped our history with an almost religious frenzy. Still, his collection grew to include some more astonishingly rare machines, such as a Boeing F4B-4 biplane fighter and a Boeing P-26A Peashooter – the latter being found still in service with the air force of Guatemala. The first home for what Maloney had named simply as The Air Museum was in the university town of Claremont. Located near the Cable-Claremont Airport in a dusty field with a large lean-to, the Museum opened for business, displaying an exotic collection of aviation hardware to anyone who was interested in paying the low admission price.

Maloney did not look back, and kept on acquiring aircraft that were threatened by the junkman. The museum moved to much larger facilities at Ontario Airport and operated successfully for several years. It was during this period that a couple of aircraft were made airworthy by enthusiastic volunteers, and the shape of the future was forged. Unfortunately, Ontario began a rapid expansion as a secondary airport to LAX, and Maloney's space was needed for future growth. With nowhere to go, Maloney hauled a number of the aircraft to a vacant lot near the field where they were heavily vandalised and, in some cases, destroyed, before acquiring a new base for some of the machines in Buena Park, where they were displayed as the Planes of Fame.

This endeavour eventually collapsed, and the majority of the aircraft found themselves at Chino during the late 1960s and early 1970s. Chino Airport was ideal for operating exotic aircraft at that time, since it was surrounded by farms and prisons – the two main industries in the area. The airport was also home to organisations such as Aero Sport, which specialised in the rebuilding of Mustangs, and it was not uncommon to visit their hangar and ramp and find around two dozen P-51s undergoing servicing. During this period Maloney's oldest son Jim and Jim's boyhood friend

At one time or another Chino has seen the majority of surviving Thunderbolts pass through its hangars. The Air Museum owns N3395G, a P-47G (USAAF s/n 42-25234) built by Curtiss at Evansville, Indiana, which is flown regularly. This aircraft was rescued by Ed Maloney from the Cal-Aero Technical Institute in Glendale when the school no longer had a use for the instructional airframe. In company with the razorback G-model is P-47D (USAAF 45-49205) N47RP (ex-N47DE) owned by Bob Pond's Planes of Fame East. The D-model is being flown by Steve Hinton while John Maloney is at the controls of the P-47G.

(Below) As mentioned in Chapter I, N47DE was flown across the Atlantic to become part of the Warbirds of Great Britain, but was rarely utilised. The aircraft was allocated the registration G-BLZW when purchased by Stephen Grey in 1985. Shortly afterwards the Thunderbolt was sold to Bob Pond; it is seen here being rebuilt at Steven Hinton's Fighter Rebuilders, which occupies some of the space at The Air Museum. The company has restored two examples of this heaviest and hardest-hitting of World War 2 fighters, among many other projects.

Steve Hinton had become hooked on the old warbirds, and began working on the aircraft in their spare time. While still teenagers both became skilled pilots and mechanics, amd much of the heritage of warbird aviation is owed to these two men.

Convinced that the museum's aircraft needed to fly, the two made several machines airworthy and inspired volunteers to do the same; and it was not all that long before The Air Museum could field an array of flying warbirds. As interest in such aircraft increased, so did the fortunes of The Air Museum. However, an economic slump during the 1970s resulted in several of the aircraft being sold to help finance the operation.

Jim's younger brother John also became deeply interested in the aircraft, and started to pitch in. In the meantime, Hinton was making a name for himself by racing at Reno in a variety of aircraft, including the highly-modified Griffon-powered RB-51 *Red Baron*. During one Reno race Hinton experienced an engine failure and crashed in a blinding explosion. The race announcer stated that Hinton was dead; but a rescue team found him still alive, though badly injured, strapped in what was left of the cockpit area. After months of recuperation, including repair of a broken back, Steve Hinton was back flying.

Fortune was less kind to Jim Maloney. While Jim and Steve were visiting an Arizona airshow during the mid-1980s both went for a hop in Ryan PT-22s flown by other pilots. No one really knows what happened; but Steve found the wreckage of the other PT-22 on the desert floor. Jim and B-17 pilot Jim Orten had died in the crash of the little trainer. Warbird aviation exacts a very high price.

As the museum prospered, more businesses which engaged in similar activities opened on the field. Leroy

Penhall opened Fighter Imports, rebuilding ex-Royal Canadian Air Force T-33s and F-86s for the civilian market. After Penhall was killed in the crash of a Beech Baron, Frank Sanders took over the large and modern hangar. Sanders, one of the gurus of warbird aviation, had rebuilt and flown several aircraft including a Mustang, P-40N and Sea Fury. He became intimately associated with the big Hawker Sea Fury and would go on to own several more examples, culminating in the magnificent, highly-modified *Dreadnought* which sported a huge R-4360 radial in place of the Bristol unit. This powerful racer ultimately conquered Reno. Frank, his wife Ruth and sons Dennis and Brian became some of the best-known personalities in warbird aviation.

Frank had become actively involved in jet restoration, and purchased a Canadair T-33 which he modified for airshow work as the *Red Knight*. Tragically, Frank and a passenger were killed when the jet suffered a catastrophic in-flight failure prior to an airshow in New Mexico during 1991. His death was a major blow to the warbird movement, but his family has successfully carried on with the business.

Other warbird organisations also call Chino home, including David Tallichet's extensive Military Aircraft Restoration facility which houses a huge collection of warplanes that Tallichet has acquired over the years. The now defunct Unlimited Aircraft Limited specialised in rebuilding not only prop-driven fighters but also the Chinese MiG-15s and other jet warbirds that arrived in the USA during the mid-1980s. The Yankee Air Corps

as built up a fabulous collection of warbirds which are
being rebuilt to flying condition, but are rarely
own. This is one of the world's most significant
ollections of really finely restored fighter aircraft and
well worth a visit.

Aero Trader specialises in rebuilding B-25 Mitchells
nd has built up a vast warehouse of hard-to-find parts.
he company, run by Carl Scholl and Tony Ritzman, is
votal in the warbird movement and has been
sponsible for a number of non-Mitchell restorations,
cluding a recently restored ex-RAF Gnat jet trainer.
acific Fighters, as the name implies, specialises in
estoring and maintaining vintage fighter aircraft, but

the company also handles a wide variety of other
warbird tasks.

Within The Air Museum, Steve Hinton, who is now
married to Ed Maloney's daughter Karen, has set up a
very successful business called Fighter Rebuilders;
their most recent triumph has been the creation of a
flyable P-38 Lightning from a pile of scrap for British
warbird collector Stephen Grey. Fighter Rebuilders
has also been responsible for several successful racing
aircraft.

The warbird enthusiast could easily spend a
weekend browsing among the various hangars and
museums at Chino, and meet lots of friendly people in
the process. Chino is one of the few airfields that has
gone from graveyard to capital of warbird preservation,
and we should all appreciate that fact.

**An interesting formation of two former
adversaries: The Air Museum's ultra-rare
Mitsubishi A6M-5 *Zero* formates on the
museum's Mitchell for a photography
session. Currently the only authentic *Zero*
flying, this fighter was one of two rescued
during the 1950s by Ed Maloney shortly
before they were due to be scrapped.**

One of the real heroes of the Pacific was the Douglas SBD Dauntless, which helped achieve early American victories at Midway and elsewhere. Unfortunately, the Dauntless is one of today's rarest warbirds since most of the survivors were scrapped after the war. During the late 1950s Ed Maloney obtained the wingless fuselage of SBD-5 BuNo 28536 and put it on display in his museum. During 1986 the decision was made to restore the aircraft to airworthy condition after a set of damaged wings had been acquired. The wings had to be completely rebuilt, and during this process it was discovered that the fuselage had served with the Royal New Zealand Air Force as NZ5062 during the war. Painted in a US Navy mid-war tri-colour camouflage scheme, Dauntless N670AM displays its classic lines over Chino. Note details such as the centreline bomb cradle and twin .30 calibre Browning air-cooled machine guns in the rear cockpit.

(Below right) Steve Hinton demonstrates the *Zero's* excellent climb performance as he departs Chino Airport. Note the landing gear coming up into the wells. N46770 was stored at the museum's several locations, but after the move to Chino a decision was made to bring this historic aircraft back into flying condition, and a complete rebuild was instigated.

(Below) Steve Hinton is equally at home in the cockpit of the *Zero* as in a jet fighter. Since parts for the Sakae radial engine are extremely difficult to come by, the condition of the engine is closely monitored and few flying hours are logged. After the restoration was completed the *Zero* made a triumphant return to Japan where it was seen by millions of spectators.

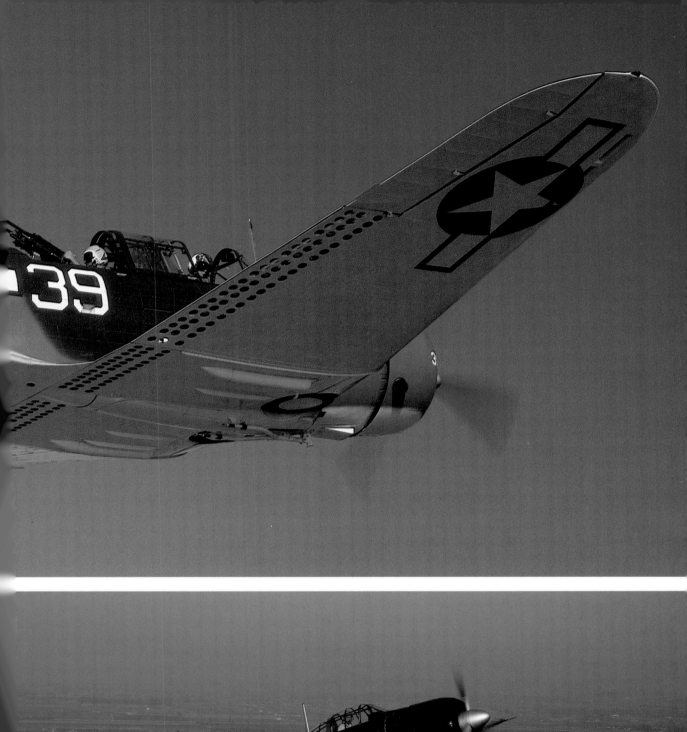

(Above) Rick Brickert pilots The Air Museum's Vought F4U-1A Corsair N83782 while Howard Pardue flies wing in his USMC-marked Goodyear-built FG-1D. The former was saved from the scrapper at the last minute when Ed Maloney rescued it from a vast aircraft storage yard owned by MGM Studios. The studio was developing the property and only last-minute notification allowed a few of the aircraft to be saved, the majority, including two B-29s, being cut up during 1970. BuNo 17799 (N83782) required extensive restoration but is now a regular flyer. Note the addition of a second seat behind the pilot.

(Below) In 1987 Steve Hinton's Fighter Rebuilders received this hulk of a P-38 from a small airfield in Texas, purchased by British collector Stephen Grey. The company was faced with the daunting (in the opinion of many, impossible) task of making the pile of corroded aluminium into a flyer. However, Fighter Rebuilders proved them wrong and s/n 42-67543 flew again during January 1992. This type of activity is not uncommon at Chino; Fighter Rebuilders is currently restoring a Bell P-63 Kingcobra and a Grumman F7F Tigercat for Bob Pond.

(Right above) Few pilots are better at close formation work than Steve Hinton, a fact amply illustrated by this close-in view of him piloting the Lockheed P-38J Lightning owned by Planes of Fame East. This aircraft had been the property of The Air Museum for many years and, when obtained by Bob Pond, was rebuilt back to flight status by Steve's Fighter Rebuilders.

(Right) Steve displays N29Q (USAAF s/n 44-23314), wearing D-Day invasion stripes, over a low cloud deck. One of the rarest of all surviving World War 2 fighters, the Lightning is also one of the most difficult to rebuild because of its complexity of construction, and the fact that the two Allison engines mean two overhauls and two cooling systems. After the war this aircraft was given to a trade school for use as a technical aid, and was saved by Maloney when the airframe became redundant. This Lightning is one of the few to retain its original fighter nose – most civilianised variants were F-5 photo-recon aircraft. N29Q divides its time between Chino and the Planes of Fame East museum in Minnesota.

Unusual aircraft are usual at Chino as they arrive to be based at the field, or for maintenance or restoration. This surplus Grumman C-1A Trader wears its original US Navy colours and is seen orbiting over Chino in company with P-51D N5441V.

One of Bob Pond's latest acquisitions for his Planes of Fame East is Supermarine Spitfire FR.Mk.XIVe N8118J, which readers might remember as being previously owned by Spenser Flack in Britain, registered G-FIRE and painted a glorious overall scarlet. This aircraft was found displayed on a garage roof in Belgium, in extremely poor condition. Flack had it completely rebuilt, and the Rolls-Royce Griffon powerplant was shipped to Dave Zeuschel in California for overhaul. The first flight took place on 14 March 1981. When sold to Pond, the Spitfire was shipped to Chino for assembly and test flying. Mike DeMarino is seen piloting the fighter near Chino.

51

Another extremely rare aircraft owned by The Air Museum is North American P-51A Mustang (USAAF s/n 43-6251) N4235Y. The Allison-powered variant of the famed fighter was, once again, rescued by Ed Maloney from a trade school; somehow the fighter had escaped scrapping after the war and was donated to Grand Central Aircraft School. The P-51A was kept at the various facilities Maloney used before becoming firmly established at Chino. In the late 1970s restoration commenced on the airframe, which was found to be in surprisingly good condition. Finished in RAF markings, N4235Y is being piloted by Rick Brickert and is accompanied by Dennis Sanders in the museum's P-51D, N5441V.

(Above) This is how many projects start at Chino: the battered remains of a Curtiss P-40E Kittyhawk recovered from the Aleutians lie forlorn in Aero Trader's storage yard, awaiting the magicians' touch.

(Right) Virtually every area of Chino has some form of warbird activity going on, whether storage or restoration. These ex-Indian Air Force *Ajeet* jets are held in store by the Military Aircraft Restoration Corp. for possible rebuild.

(Below) An exciting project at The Air Museum is the restoration of a Japanese Aichi *Val* dive bomber, acquired from Canada in poor condition; museum volunteers have had to completely rebuild many sections of the rare aircraft. It is expected that a first flight might take place by the end of 1992. The powerplant will be an American engine due to the unavailability of the original Japanese radial.

53

4: Museum of Flying

Some of the North American T-28 Trojans based at Clover Field are seen out over the Malibu coastline during one of the regular sessions undertaken to keep up formation skills – one of the most neglected elements in general aviation flying. The Museum of Flying's Trojan is the all-yellow example finished in accurate USMC instrument trainer markings; it was donated to the MOF by Chuck Smith.

The oldest airport still operating in southern California is Clover Field, located atop a mild plateau in the city of Santa Monica. Established in 1918 and named after a World War 1 aviator, the airport has been operational ever since. Those early years saw the dirt and grass field filled with a variety of surplus ex-World War 1 aircraft; but southern California was rapidly becoming the home for the new aviation industry, and many new types of aircraft began testing their wings over Clover Field. The mild climate of the area attracted many fledgling aeronautical concerns – though most of these companies came and went fairly quickly, since aviation can be, at best, fickle.

During 1920 a young aeronautical engineer named Donald Douglas left his job at an east coast aircraft manufacturer and moved to Santa Monica. With only a limited aeronautical background and $600 in capital, the young man attempted to set up an aviation concern in the area which he felt would eventually be the main centre for international aviation activities.

From beginnings in the back of a barbershop on Pico Boulevard, Douglas rose to greatness. Subsequently building his designs on a bankrupt film company's lot on Wilshire Boulevard, he had his airframes towed two miles to Clover Field for flight testing. His first great triumph came in 1925: four US Air Service Douglas World Cruisers left Clover Field on 17 March, and 175 days later three of them arrived in Seattle, having circumnavigated the globe and earned Douglas the proud logo 'First Around The World'.

In 1929, reorganised as the new Douglas Aircraft Company Inc., the firm moved into much larger facilities at Clover Field. From that day on there was no stopping them. More modern factories at Long Beach and in Oklahoma City would rise in time; although Douglas continued building DC-4/C-54 transports at Santa Monica after the war, the 5,000 foot runway was a limitation. Most manufacturing was transferred to Long Beach, although components and missile systems were still made at Santa Monica until the early 1970s. At last the famous plant was closed down; and, sadly, a preservation order on the immense wooden building –

One of the more significant aircraft acquired by the museum during its formative stages was Supermarine Spitfire FR.XIVE NH749, purchased by David Price in Britain during 1985. In this photograph the new owner is seen in NH749 during a sparkling clear day off Malibu as he leads Alan Preston in Tiger Destefani's Kittyhawk, both in World War 2 Commonwealth markings. Federal Aviation Administration bureaucrats decided the RAF serial on the Spitfire looked too much like an American civil registration and ordered the N removed; the actual civil registration is NX749DP. Last used by the Indian Air Force, it was one of a group of Spitfires found derelict by the late Ormond Haydon-Baillie during 1977. Seven aircraft were shipped to Britain; after Haydon-Baillie's death in a Mustang the fighters were sold, and NH749 went to A. and K.Wickenden.

claimed to be the longest contiguous wooden structure in the country – did not stop developers from bulldozing it in the late 1970s.

In the place of the classic factory, a modern phoenix rose under the guidance of international businessman David Price. Price, a former US Navy aviator, envisioned a modern aviation business complex that would include a fixed base operator to service business aircraft, a large hangar area, maintenance facilities, office space, a famous restaurant – and a world-class aviation museum serving to document the achievements of Donald Douglas and other pioneers in what had become known as southern California's cradle of aviation.

Accordingly, a beautiful modernist four-storey structure was erected that would become known as The Museum of Flying. The centrepiece of the museum was one of the surviving Douglas World Cruisers: hung from the rafters as if in flight, the aircraft was a major coup for the young museum, which opened its doors in 1989 and has not looked back. One of Price's goals

was to make the organisation a flying museum, including regularly operating veteran and vintage aircraft. The museum would encompass a growing collection of flying aircraft that would educate the public about the importance of aviation, and how flying helped to shape southern California. As can be seen from the photos in this chapter, the Museum of Flying has done an admirable job.

As an offshoot to the museum, it was decided that during 1990 an auction would be held that would include the finest in veteran and vintage aircraft. A two-day auction was staged during May of that year which attracted world-wide interest and began to set a value pattern for these rare aircraft. In October 1991 a second auction was held, and despite the world-wide recession the event was a success, paving the way for an annual event. In just a few short years, the Museum of Flying has established itself as one of the premier forces in the world of warbird aviation.

(Above) NH749 was restored by Craig Charleston for the Wickendens, making its first flight as G-MXIV in 1983; however, Keith Wickenden was killed in the crash of a De Havilland Dove, and the Spitfire was sold to David Price in 1985. It is painted in British late war South-East Asia theatre markings.

(Below) The MOF's Restoration Facility director is Bruce Lockwood who is also responsible for maintaining the museum's fleet of flying aircraft. Lockwood is seen making a last minute adjustment to Mustang N151DP before a flight.

(Right) The two Spitfires on an outing over the Pacific Ocean near Catalina Island. As this book was going to press the Museum of Flying also took delivery of fully-restored Hawker Hurricane G-ORGI, and this even rarer WW2 veteran was being assembled at Chino by Spitfire rebuilder Craig Charleston for its first flight in the States.

The second Supermarine fighter acquired by the MOF is the very historic Spitfire HF.IXC/E MA793, bought from a South African owner. Registered NX93OLB, it is seen here flown by David Price during January 1989. As can be seen, the South African insignia was very similar to the RAF's except for the replacement of the red by orange.

The Museum of Flying has held two successful vintage and veteran aircraft auctions (in 1990 and 1991), and the event has become an annual gathering of international buyers and sellers. Many interesting aircraft have gone up on the auction block, including this Percival Prentice which was sold to the American Aeronautical Foundation. N1041P is the only example of its type in America.

One of the rarest aircraft to be featured in the auction was this antique Stinson tri-motor finished in the markings of American Airlines. N11153 is one of only two surviving examples of the type.

Museum director David Price is also an air racing enthusiast; here he is seen piloting Mustang N541OV *Dago Red* in company with a Blue Angels F/A-18A Hornet during a visit to the Blues' winter training camp at NAF El Centro on 29 February 1992. In the fall of 1992 the MOF will be unveiling a major air racing display.

(Below left) During 1991 the MOF took on some challenging restoration projects, including several Japanese warplanes recovered from an island off Indonesia. The hulks of a number of Mitsubishi A6M *Zeros* were acquired and, to facilitate restoration, MOF members custom-built jigs and fixtures for the aircraft. In just a few months they had made amazing progress, as can be seen by this nearly complete rear fuselage. The MOF hopes to have a flying *Zero* in three years, and will also tackle the daunting job of restoring the original engine to flying condition.

(Below) These two Folland Gnats were residents of the MOF ramp for several months in early 1991 after being the film stars in '*Hot Shots*', painted in bogus US Navy markings. The watercolour paint has already started to weather, revealing the Royal Air Force markings underneath.

Certainly one of the most interesting aircraft to appear at the 1991 auction was the unique Bird Innovator – a Consolidated PBY-5A Catalina that had been converted into the ultimate air yacht. Two Lycoming engines had been added outboard for additional reliability and power. Note the clear blister (with a sleeper couch on the inside), offering an exceptional view while in flight. Two large powerboats can be carried under the wings and lowered into the water for transportation to the shore. Oddly, this rare aircraft failed to find a buyer at the 1991 auction.

(Above) The Stout Busmaster is a modernised version of the famous Ford Tri-Motor; only two were made, one in the 1960s, the other during the 1980s. The second example, N75ORW, was sold at the 1990 auction to California Wings Air Tours, which uses the modernised classic for sightseeing tours.

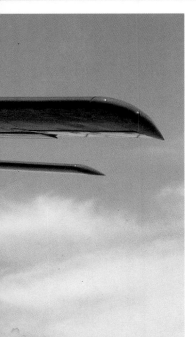

The only surviving Helio Stallion was offered for sale at the 1991 auction and quickly found a buyer. This beautifully restored example was originally used by the CIA in South-East Asia for covert warfare; capable of operating out of short and primitive strips, the Helio Stallion could carry a variety of weapons.

(Left) A small selection of the 1991 auction offering is seen displayed on the MOF ramp.

(Above) A real powerhouse, and one of the more attractive aircraft at the 1991 auction, was this modified Yak-11 owned and flown by airline captain Joe Haley. Joe replaced the aircraft's Soviet engine with a Pratt & Whitney R-2000 that greatly increased the horsepower, and he has raced the Yak, named *The Defector*, at Reno.

(Below) In aviation there's a market for just about anything. . . . This wrecked Lockheed F-5G Lightning was discovered in the California desert by Joel Bishop and put in the 1991 auction. It was quickly purchased by famed Florida rebuilder Tom Reilly, who vowed that the aircraft, when combined with other parts, will eventually fly again.

Straight from winning a top award at the 1991 Oshkosh fly-in, this finely restored Twin Beech in US Navy utility squadron colours was quickly bought at the auction for a record price. Eric Clifford had done a complete rebuild of the aircraft, which is a rare photographic variant with camera windows in the belly, to create a mint example of one of Beech's most famous products.

The star of the first MOF auction was Lockheed P-38L/F-5G Lightning N5596V, USAAF s/n 44-26981. This aircraft, when fitted with its camera nose, was used by several photo-recon companies after being surplused from military service. The gleaming Lightning emerges from the Aeroflair paint shop in full D-Day markings for the 20th Fighter Group, the photo-recon nose replaced with a handbuilt fighter unit manufactured to original specifications by Phil Greenburg. The twin-boom fighter is one of the rarest World War 2 aircraft in the skies, with only about six examples still airworthy. N5596V was heavily damaged in a landing accident during 1972 at Paris, Texas; considered a write-of, it went through a couple of owners before being obtained by John Silberman of Florida, who was determined that the fighter would fly once again. He enlisted the help of Tom Crevasse, and after a long and arduous battle the Lightning was finally pieced back into flying condition. After further rebuild work at Santa Monica the P-38 was sold at the 1990 MOF auction to retired USAF General William Lyons, a former World War 2 Lightning pilot, for a record $1.5 million.

On 23 December 1991 the MOF received a donation of the last flying Armstrong Whitworth Argosy C.Mk.1 courtesy of Harry Barr and Duncan Aviation in Lincoln, Nebraska. The big four-engine turboprop transport had been operated in Alaska on Bureau of Land Management contracts since the mid-1970s. The museum escorted the aircraft back to Santa Monica in fine style with Mustang N151DP and newly painted Spitfire Mk.IX N93OLB. The massive Argosy was used as a heavy-lift transport by the Royal Air Force, and will be maintained by the MOF in operational condition, although it will only be flown on special occasions.

Bruce Lockwood took the Spitfire Mk.IX up for a test flight on 31 December 1991 and had a decline in oil pressure and a runaway propeller. Out over the Simi Valley, Bruce fortunately managed to find an abandoned airstrip and deadsticked the fighter onto a disused runway. The Spitfire was disassembled and trucked back to Santa Monica for repairs.

The MOF's P-51D N151DP, with Alan Preston piloting, is seen out over the Pacific in company with Bill Rheinschild's P-51D N5415V, flown here by Bruce Lockwood. The 'Southern Cross' rudder marking identifies an Australian unit flying over Italy and Yugoslavia in 1944 with 239 Wing of the RAF's Desert Air Force.

5: Trainers, Transports and Oddities

The T-28B was created out of a desire to standardise USAF/USN training aircraft. The B-model was developed for the Navy and featured the more powerful Wright R-1820-86 engine of 1,425 hp, a new prop and other changes to suit the aircraft for the Navy's way of training. The first B-model flew on 6 April 1953, and North American went on to build 489 T-28Bs. The Navy phased the type out in 1984, and the Trojan has been in heavy demand among warbird collectors ever since. T-28B NX57973 (ex-USN BuNo 137801) is attractively finished in the markings of VA-122, an attack squadron that actually did have a few Trojans on strength. Trojans have been in demand as counter-insurgency aircraft (one of the most recent uses was by dissident Philippine Air Force pilots attacking President Aquino's palace in their all-black T-28Ds in December 1989); but heaviest demand comes from American private pilots, who search far and wide to discover foreign airframes that can be brought back to the USA and restored.

During World War 2, California became a vast training base for all types of military activity. The state's huge agricultural-based Central Valley sprouted dozens of training fields and the air soon hummed with the sound of PT-17 Kaydets, BT-13 Valiants, and AT-6 Texans as tens of thousands of young Allied pilots began to earn their wings. The climate was such that training could go on all year.

After primary training the pilots went on to other bases – once again, many were in California – to allow students to complete training in bombers, transports or fighters as quickly as possible. Some bases had specialised training, such as Hammer Field in Fresno, which concentrated on producing night fighter crews;

and Blythe, in the desert on the Colorado River border between California and Arizona, which turned out complete B-24 Liberator crews.

Southern California had rapidly become the world's aviation arsenal, where companies like Lockheed, Northrop, Douglas, Vultee, Consolidated and North American were churning out thousands of high-quality combat aircraft. During the war years California became established as the world's aviation capital – a surprising feat, considering that the state's main occupation before the war had been agriculture. The rapid build-up of arms also meant a massive influx of population, both military personnel and the workers for the war plants. After VJ Day, many of these

(Right) An aeroplane that survived the mission for which it was constructed! Culver Aircraft Company built a series of popular light aircraft for the civilian market before World War 2. During the war the company designed an expendable all-wood target drone designated PQ-14, and went on to build nearly 2,000 of them – the majority powered by the Franklin 0-300 engine. Only a few survived the war, but some were put on the civil market as single-seat aircraft, and they made a rather nice sport plane. Today only a couple still fly, including the all-red NL51HM, owned by The Air Museum and seen near Chino being flown on a rare outing by Robbie Patterson.

(Left) Certainly one of the most popular 'big' trainers, the North American T-28 Trojan is particularly in favour among pilots who do not have 'taildragger' time. Originally built for a USAF competition to create a replacement for the classic T-6 Texan, the NAA Model 159 combined primary and basic training characteristics in a single craft, thus reducing the number of types in military service with a consequent monetary saving. Alan Prestor is seen flying lead in a T-28A airframe NX5051L, which has been fitted with the much more powerful Wright R-1829-56S radial – similar to that installed in the combat version of the Trojan, the T-28D (1,425 hp): the
A-model's original Wright R-1300 engine produced 800 hp, while the '-56S pumps out a significantly increased 1,300 hp. Bruce Lockwood flies wing in Trojan N1742R, also modified with the uprated engine.

individuals decided that they liked California and stayed, swelling the state's population.

Although politicians had predicted an idyllic life after the successful conclusion of the war, actuality reared its ugly head: there were thousands of layoffs and a recession as military contracts were either cancelled outright or severely cut back. There had also been predictions of 'an aeroplane in every garage' to satisfy the demands of the thousands of aviators returning from the war. The sad fact was that many of these pilots never wanted to see an aeroplane again – they had fulfilled their war duties and now wanted to get on to other things, like raising a family, buying a house and getting an occupation. The companies that had built a variety of small aircraft designs to cater to this market quickly fell upon hard times.

The RFC

However, for military pilots who did want to keep on flying, and for their civilian counterparts who wanted to get back into the air after the lifting of the wartime ban on flying, there were some amazing bargains. The Reconstruction Finance Corporation had taken over responsibility for disposing of America's vast fleet of surplus military aircraft. Although the programme was plagued by inefficiency and scandal (for example, some scrappers were able to buy hundreds of B-17s, drain the high-quality avgas from their tanks, sell the fuel to pay for the purchase price of the aircraft, and

then scrap the aircraft at an excellent profit), the RFC managed to sell off the fleet in only a couple of years. Fortunately, civilians could also buy these aircraft; pilots got low-time training aircraft for just a few hundred dollars, while veterans who dreamed of starting their own airline were able to buy C-47s, C-46s and C-54s – although at considerably higher prices. Some companies needed surplus bombers for specialised duties such as extensive conversion or high-altitude mapping. Others wanted the speed of a fighter – some for the famed Cleveland National Air Races that had started again after the war, some for more clandestine operations such as smuggling the aircraft to Third World nations.

Ag Flying

Although the majority of America's combat fleet was scrapped, some did survive, especially the trainers. In 1946 California was filled with small 'mom and pop' type airfields (most now, sadly, gone because of increasing property values), and these strips were soon alive with hundreds of surplus military training aircraft. Some of the larger trainers, like the Cessna UC-78 'Bamboo Bomber', were suitable for passenger charter work. The growing agricultural application industry snapped up Stearmans by the hundred, and the Central Valley once again echoed to the sound of training aircraft, although now on a vastly different mission.

As with most ag flying, there was always a need for

Everything down and out, Jeff Kertes sets up a T-28A N28NA (ex-USAF 49-1630) for the camera to show what the Trojan looks like in landing configuration. Owned by the Cactus Air Force, this aircraft is finished in the particularly colourful paint scheme carried by a T-28A assigned to Edwards Air Force Base during the late 1950s. Note the three-blade propellor modification.

more power, and the dusters and sprayers soon discovered that vast numbers of Vultee BT-13 Valiants could be picked up very cheaply. The operators flew or hauled the BTs to their fields where the engine and prop were removed and added to a Stearman, thus greatly increasing the aircraft's power and load-hauling capability. The BTs, which now had little or no value, were simply left to rot; into the 1970s it was still possible to visit duster fields and find these hulks quietly corroding away. Fortunately, most of what was left was scooped up by the end of the decade as restorations of BT-13s and other trainers became popular and profitable.

After several decades of very hard use, the Stearman fleet was phased out of service in the Central Valley in favour of newer equipment. Fortunately, the majority of the survivors went to individuals or companies who have painstakingly restored the trainers to their original military configuration. Stearmans have always been popular airshow mounts, usually equipped with bigger engines and smoke systems, and the type still remains in healthy demand for airshow performances.

The Airlines

After the war Oakland and Burbank Airports became the homes for dozens of new non-scheduled airlines – usually started by a few ex-military pilots who got together to pool their mustering-out pay and buy a transport, with visions of growing into multi-plane operators. These 'non-skeds' would go anywhere and haul anything, and were greatly undercutting the regular airlines, causing consternation and heavy political lobbying. Dozens of new outfits, many having only one aircraft, began operating from these airports.

creating a vibrant galaxy of names and colour schemes. These brave little enterprises would go in and out of business with great regularity, but it seemed as if there was always another hopeful popping up to fill a gap. After a few years the political lobbying paid off and the 'majors' were able to force most of the 'non-skeds' out of business, thus ending a brief but colourful chapter in aviation history.

California has, rightly or wrongly, acquired the reputation of being the land of 'fruit and nuts', and this philosophy can, unfortunately, sometimes be applied to the aviation community. What person in their right mind would purchase a surplus military aircraft? These aeroplanes are often difficult to license, dangerous to fly and extremely expensive to operate. Yet, curiously, after the war a number of rather 'eclectic' aviation-related individuals acquired large fleets of these

phased-out warriors. What their original intention was is now, with the passage of time, somewhat unclear; but it remains a fact that due to their now-forgotten schemes a number of California airfields were littered with parked ex-military airframes ranging from Corsairs to Mitchells. That some of these aircraft survived into the 1970s is one of the reasons that today's warbird movement is so strong. After these individuals either died or faded from the scene their collections were sold off to form the basis of today's warbird scene. Many strange and wonderful 'one of a kind' aircraft were brought back to life and are now a regular part of the Californian airshows; and some of these machines are illustrated in this chapter.

Brian Sanders is seen flying his family's Beech T-34 Mentor on an unusually clear day out of Chino. Frank and Ruth Sanders, along with sons Dennis and Brian, established a successful warbird rebuilding and modification business at Chino which has created many fine aircraft, including the Sea Fury racer *Dreadnought*. After Frank's tragic death at the controls of the *Red Knight* T-33 in 1990 the family continued the business; one of their most recent projects to be completed is a magnificent Commonwealth Aircraft Corporation 'Boomerang' fighter. Currently, many T-34s operate in

California; it is an aircraft in high demand, since it has a nosewheel undercarriage and can be easily flown by pilots without 'taildragger time'. Based on the Beech Bonanza, the T-34 was produced as a tandem military trainer for the US Air Force and Navy. N311H, which has since been sold to a new owner, was a stock airframe when obtained by Frank, but he replaced the original 225 hp engine and two-blade propellor with a Continental 285 hp powerplant and three-blade prop which greatly increased performance. N311H was used as a camera platform for some of the photographs in this book.

(Below) Ex-military jets operated by civilian owners are a small but growing part of the California warbird scene. The numbers have greatly increased with the infusion of ex-Warsaw Pact or ex-Soviet jets imported following the collapse of the Communist system: former Communists have quickly turned capitalist and many jets have been released for a transfer of 'greenbacks'. One of the most attractive is the Yugoslavian Soko Galeb, offering a tried-and-true Rolls-Royce Viper powerplant and modern systems, along with rugged construction. This attractively finished example is seen over Lake Tahoe.

Unusual aircraft are hardly strangers to the skies over southern California, but certainly one of the most interesting visited us during January 1991 when a vintage *Lufthansa* Junkers Ju 52/3m droned its stately way through the Los Angeles airspace. The aircraft (finished as D-AQUI, but actually registered D-CDLH) had been airlifted into Canada from Germany by one of the Soviets' huge transports (a goodwill gesture), and then assembled and flown to the United States. The aircraft was on an ambitious tour, visiting the various cities that *Lufthansa* services and giving rides to employees and friends of the airline. We were able to meet up with the Ju near Los Angeles International Airport, and during the photo-sortie we had a serendipitous encounter with the Goodyear Blimp.

This particular aircraft had seen a long, hard life, operating at one time on floats in Norway, before ending its days in Ecuador where it was discovered by American crop duster Lester Weaver. Weaver patched the tired Junkers back into flying shape for its lengthy trip north; and after flying it for several years, sold the aircraft to the well-known aviation writer Martin Caidin, who invested a considerable sum restoring it to airshow shape. Caidin had great fun with his old transport; but *Lufthansa* made an offer he could not refuse, and in due course the old lady was flown home across the Atlantic to Germany, where it was restored to pristine condition by

Lufthansa volunteers. The final result was a splendid example of one of the world's pivotal passenger and military transport aircraft. *Lufthansa* campaigned the aeroplane in Europe for several years before making the decision to bring it to the United States.

(Above) One of the least successful observation aircraft built for the US Army Air Corps, the Curtiss 0-52 Owl was an interesting machine with a number of innovative features. Curtiss designers drew heavily on their earlier naval experience, using a retractable landing gear similar to the BF2C-1, while a collapsible rear turtledeck structure gave the rear gunner a clear field of fire with his single .50 calibre machine gun (similar to the solution adopted on the SOC and SO3C Seagull). The R-1340-51 of 600 hp was chosen for the powerplant: but a combination of drag, heavy weight and an inefficient wing left the 0-52 distinctly underpowered, with an almost non-existent rate of climb. The Air Corps procured 203 Owls, but the performance was so dismal that the aircraft were not issued to frontline units and went directly into a training role. After the war several were purchased surplus and used for mapping. This example was bought by the Yankee Air Corps at Chino, and was extensively restored to original 1940 condition complete with all military equipment. Finished in period Olive Drab and Neutral Grey, the aircraft is seen during one of its very rare aerial outings piloted by Gary Meermans.

(Right above) Never say never. . . . During the Reagan years it was virtually an impossible thought that ex-Soviet military aircraft would operate in the skies of California. However, after the collapse of the Communist bloc what was thought impossible became, if not exactly common, a regular sight at airshows. One of the first outfits to receive some ex-Soviet hardware was the ever-innovative Planes of Fame Air Museum at Chino: the museum received a massive Antonov AN-2 biplane, and quickly got the beast assembled and flying. A true 'sky truck',

the AN-2 has to be one of the most efficient aircraft ever built (and it was, and apparently still is, built in huge numbers), serving in military roles, as a bush plane, crop duster and airliner, to name but a few occupations. The distinctive 'flying cathedral' lines of the AN-2 are seen droning along in the Chino pattern under the capable control of Don Lykins. Registered NX90490, the AN-2 is an extremely popular airshow attraction and is one of about a half-dozen currently flying in the States.

(Right) What aircraft could more typify the classic primary trainer of World War 2 than the Boeing Stearman Kaydet? A 1936 contract would eventually see well over 8,000 constructed for America and its allies. Stearman was taken over by Boeing, and a variety of Kaydets were constructed including the PT-13, -17, -18, -27, N2S-1, -2, -3, -4, and -5. Most of these redesignations involved minor changes with the addition of Lycoming or Continental engines, but full Army/Navy interchangeability was finally achieved with PT-13D/N2S-5. After the war thousands of airframes were cheaply surplused to pilots who wanted a fun flyer, or to business concerns who wanted a sturdy airframe for the rapidly expanding role of aerial crop application. During the 1950s, 1960s and early 1970s

hundreds of Stearmans could be seen flying from dozens of small airfields in California's Central Valley as they waged war with the bugs. However, as the years went by the Kaydets became less economical to maintain, and newer aircraft capable of carrying much larger payloads took their place. Most of the Kaydet fleet were not junked, but fell into the loving hands of restorers who recreated the aircraft in their original military configuration. Today, several companies specialise in the recreation of Kaydets and the aircraft is one of the most popular of the ex-military trainers, since items like new-built wood wings are easily obtainable. N56226, attractively restored in US Navy markings, is seen over its post-war 'battleground'.

One of the toughest 'L-Birds' to operate with American forces during World War 2 was the Stinson L-5 Sentinel. Developed from the pre-war civil 105 Voyager series, the aircraft obtained by the Army was greatly strengthened and heavier, although it still used the previous aircraft's 34 ft span wings and 0-435-1 engine of 185 hp. Delivered in a number of variants (including an ambulance version), over 3,000 Sentinels were constructed for military use. After the war a great many were surplused to civilian owners but, as the years wore on, were usually left to rot in the airport weeds – victims of 'too expensive' engine overhauls, fabric re-covering, etc. However, the recent interest in restoring and flying the 'warbugs' (a very appropriate name penned by Budd Davisson) has caused a great many 'basket cases' to be salvaged and brought back to life: currently, around 30 Sentinels are flying and many more are under restoration. N69892 is a magnificently restored example of the type and is seen in full D-Day markings, along with some wonderful nose art, above Madera, California, during the annual Gathering of Warbirds airshow.

Certainly one of the *worst* sights in the California sky was this incredibly mangey Norduyn UC-64 Norseman, registered N1037Z. Saved from an aircraft junkyard by well-known aeronautical horse trader Ascher Ward, the aircraft – which had last seen service with a Canadian fishing camp – was patched up into (barely) flying condition and flown to Van Nuys, California, for further work. Designed in Canada as an efficient pre-war bush plane, the type was adopted by the USAAF in 1942 and 746 were purchased. Flown by Matt Jackson over the Santa Monica mountains, N1037Z displays an impressive number of patches on its scabby fabric hide.

(Right above) The PZL Wilga makes for a strange sight in any sky, but especially over California, where the type is arguably more alien than a flying saucer. Built as a solid utility workhorse, the Wilga is employed in a number of variants in its home country, from crop sprayer to glider tug. One of a small number imported over the years, this machine is attractively, if inaccurately, finished in Polish Air Force markings. Note the large cabin and excellent visibility, the large steps, and the glider tow attachment on the tailwheel.

(Right) California is home to virtually all the world-class air racers that make their yearly appearances at the Reno National Air Races in neighbouring Nevada. Speed, whether with cars or aircraft, has always been a Californian tradition; and it therefore came as little surprise when air racing enthusiasts wanted to strap a massive Wright R-3350 turbo-compound radial engine onto the relatively tiny airframe of an ex-Egyptian Air Force Yak-11 trainer. The Yak had the same small wing as the famous World War 2 Yak-3 fighter – which, with the huge and heavy engine, meant high wing loading. The Yak-11 airframe naturally had to be heavily modified, and a team led by Matt Jackson and Pete Regina extensively reworked the fuselage to incorporate extremely heavy tubing and the additional cooling system needed for the big radial. With intrepid test pilot Skip Holm strapped into the cockpit

of N134JK, the highly-modified Yak is seen on its fourth test flight, the nose-high attitude betraying Skip's attempt to slow the beast down to maintain formation with the Texan camera plane flown all-out by Jack Ward. After further testing it was decided that the racer needed more tail surface, and the horizontal and vertical from a T-33 were grafted onto the aircraft. This did not, however, create a bright future for the racer. By the time of the 1988 Reno event ownership had passed to famed race pilot Darryl Greenamyer, who hired former General Dynamics test pilot Neil Anderson to fly the aircraft. Neil suffered an engine failure shortly after take-off, and the Yak was virtually demolished during a crash-landing. Fortunately Neil was extracted with little more than heavy bruising; and the latest word is that the Yak is being rebuilt once again, this time with 'improvements'!

In terms of physical size the largest production twin-engined aircraft built during World War 2 was the imposing Curtiss C-46 Commando, spanning just over 108 ft and powered by two Pratt & Whitney R-2800-51 radials of over 2,000 hp each. Most surplus C-46s gravitated to Latin America where their excellent load-hauling capabilities made (and still make) them very desirable aircraft. However, a couple of Commandos now fly the airshow circuit in the USA; one of them is *China Doll*, operated by the Southern California Wing of the Confederate Air Force out of Camarillo Airport in southern California. The aircraft is labour-intensive, but a dedicated crew of volunteers keep the aerial behemoth in fine flying condition. This close-up of *China Doll* shows the whale-like lines of the fuselage, with the windshield and the fuselage contours merging into an aerodynamic shape. The art work for *China Doll* was executed by Tony Starcer, an 8th Air Force veteran who also painted the nose art for the famous B-17 *Memphis Belle* and a number of other well-known bombers. Unfortunately, Starcer passed away shortly after completing *China Doll*.